Youssef EL GUAMRI
Omar AMAHMID
Driss BELGHYTI

GASTROINTESTINAL PARASITES

Youssef EL GUAMRI
Omar AMAHMID
Driss BELGHYTI

GASTROINTESTINAL PARASITES

Epidemiological and clinical profile of children

ScienciaScripts

Cover image: www.ingimage.com

This book is a translation from the original published under ISBN 978-620-6-72209-0.

Publisher:
Sciencia Scripts
is a trademark of
Dodo Books Indian Ocean Ltd. and OmniScriptum S.R.L publishing group

120 High Road, East Finchley, London, N2 9ED, United Kingdom
Str. Armeneasca 28/1, office 1, Chisinau MD-2012, Republic of Moldova, Europe
Printed at: see last page
ISBN: 978-620-8-25437-7

GASTROINTESTINAL PARASITES

EPIDEMIOLOGICAL AND CLINICAL PROFILE OF CHILDREN

PROF. YOUSSEF EL GUAMRI PROF. OMAR AMAHMID

PROF. DRISS BELGHYTI

SUMMARY

A prospective epidemiological survey of intestinal parasitism was carried out in the Paediatrics and Child Surgery departments of the El Idrissi Regional Hospital in Kénitra (Gharb) among 300 children hospitalised in 2007. This survey identified 11 intestinal parasites: Entamoeba coli, Entamoeba histolytica/dispar, Endolimax nana, Giardia intestinalis, Iodamoeba butchilii, Blastocystis hominis, Ascaris lumbricoides, Trichuris trichuira, Enterobius vermicularis, Taenia saginata and Hymenolepis nana. Of the 300 children, 157 had at least one intestinal parasite (52.7%), with 91 cases of polyparasitism (30.3%). Among the intestinal parasites identified, oxyurosis was the most prevalent intestinal parasitosis among hospitalised children. The survey showed that the age group most exposed to intestinal parasites is between one and nine years old. The deterioration in hygiene and environmental conditions has led to a remarkable increase in the rates of intestinal parasites in children living in rural areas. Most of the parasites identified are the result of poor hygiene, which means that those who carry them are at risk.

Key words: *survey, epidemiology, intestinal parasitism, children, Kénitra, Morocco.*

I. INTRODUCTION

The quality of the environment has long been recognised as one of the main factors determining people's state of health. The new concept of environmental health is the result of the convergence of environmental degradation, public awareness of this degradation, and knowledge and progress in scientific research into the links between health and the environment (HAUET, 2004).

Over the last few decades, urban areas have grown considerably, generating ever-increasing volumes of wastewater (AHN and CHOI, 2004). Poor water quality causes diarrhoea, responsible for the death of 1.8 million people every year, including 1.6 million children under the age of five. It is also the cause of many diseases, including cholera, dysentery, typhoid and intestinal parasites (WHO, 2003).

At this level, to be effective, health protection measures require regular monitoring of the population exposed to the risk. The assessment of this health risk is carried out by epidemiological studies in which the epidemiologist will investigate the circumstances in which a disease develops in a population at risk. Furthermore, it is desirable to limit these studies to diseases linked to untreated wastewater. These are generally intestinal helminthiasis and diarrhoeal diseases (WHO, 1991).

Intestinal parasitosis is one of the most common infestations in the world. They represent a major public health risk, with a high prevalence in many developing countries. But the real problem with these parasitic diseases lies in the large number of asymptomatic carriers who contribute to the spread and perpetuation of these diseases. Infestations caused by parasitic helminths are among the most common - and most neglected - diseases, affecting more than 30% of the world's population. These helminthiases are the cause of very high mortality in developing countries (WHO, 1996), and they reflect a mesological constraint aggravated by the absence of faecal hygiene, the poor or total absence of general liquid sanitation measures and the lack of drinking water (NOZAIS, 1998).

In many countries, these intestinal parasites cause serious health and social problems, leading to malabsorption, diarrhoea, blood loss, impaired work capacity and slow growth (WHO, 1997 and GOLVAN, 1990). Given the major impact of these intestinal parasites on human health, epidemiological studies of these diseases are strongly recommended. by the World Health Organisation to establish a regular surveillance system and a programme to combat these intestinal parasites, especially in children. Intestinal parasitism is a classic phenomenon in children and plays an important role in childhood pathology as a whole, particularly in developing countries. Because of the often precarious climatic and hygienic conditions, intestinal parasitosis plays an equally important role in parasitic diseases as a whole (LAAMRANI et al., 1999).

The prevalence of different species of parasites varies from one region to another. Intestinal parasites spread rapidly in areas where sanitary facilities are inadequate or non-existent and where there is overcrowding. The corollary is an increase in faecal peril, which is seen as a consequence of the lack of sanitary facilities for collecting faeces (open dumping), the lack of drinking water and the lack of individual and collective hygiene.

The extent of intestinal parasitosis in the hospital environment is still poorly known in our country. This study was carried out in the Paediatrics and Child Surgery Department of the El Idrissi Hospital in Kénitra. We wanted to know the specific distribution in children in the Kénitra region and we report the first study concerning 300 stool examinations analysed in the Biology & Health Laboratory of the Faculty of Sciences and in the Medical Analysis Laboratory of the El Idrissi Hospital in Kénitra.

The study covered children with non-parasitic illnesses hospitalised between June and October 2007. The children included in the survey were aged between 1 month and 16 years, and had a complete medical file. This was a prospective cross-sectional survey. The complete medical file included an interview, physical examination, results of paraclinical tests, treatment received by the patient and evolution.

The results obtained were communicated only to the parents of the children hospitalised, via the doctors responsible for monitoring the state of health of the children examined.

At the end of the study, the children whose analyses revealed the presence of intestinal parasitism were referred to paediatricians for appropriate medical care.

II. ENVIRONMENT & METHODOLOGY

II.1. Study environment

The province of Kénitra is subdivided into 6 urban communes (Kénitra Maâmora, Kénitra Saknia, Mehdiya, Sidi Yahia, Sidi Slimane and Souk El Arbaa du Gharb); 4 circles (Kénitra banlieue, Sidi Slimane, Souk El Arbaa du Gharb and Souk Telt El Gharb) and 27 rural communes. In terms of public health infrastructure, the province of Kénitra has three hospitals, including a regional hospital (Hôpital El Idrissi) in Kénitra and two health polyclinics (Sidi Slimane and Souk El Arbâa) (Figure 1).

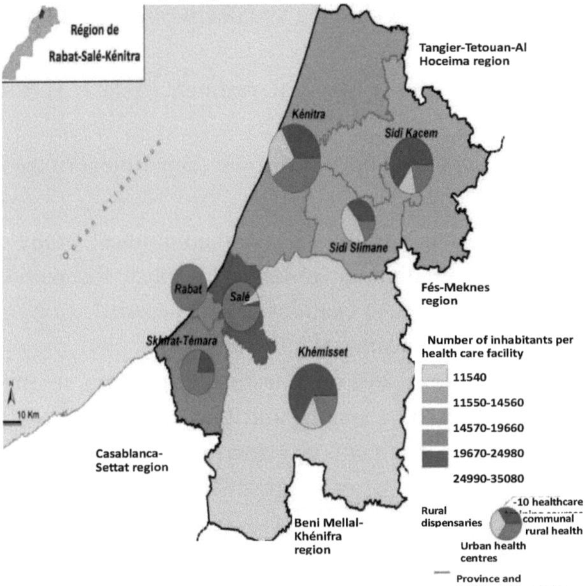

Figure 1: Health divisions by commune and cercle of Kénitra (MOROCCO)

According to the latest general population and housing census in 2004, the population of the province of Kénitra is estimated at 1,167,000, of which 572,700 (42.46%) live in urban areas and 594,601 (57.69%) in rural areas. The population is made up of almost as many men (49.9%) as women (51.1%).

This is a very young population, with 32.5% aged under 15, 43.9% aged under 20 and 62.6% aged under 30, making up more than 48.4% of the rural population compared with 29.7% in urban areas. The prospective epidemiological survey of intestinal parasitism in hospitalised children was carried out in the paediatric and child surgery departments of the Kénitra Regional Hospital Centre.

II.2. Study methods

II.2.1. Children surveyed

We wanted to know the specific distribution in children in the Gharb region and we report the first study concerning 300 children not suspected of intestinal parasitosis and hospitalised at the Centre Hospitalier Régional du Gharb (Kénitra) for another pathology. The study was carried out between June 2007 and October 2007.

The stools examined were from children aged between 1 month and 16 years.

They included :

- children hospitalised in the paediatrics department of the C.H.R. de Kénitra (Gharb),
- children hospitalised in the Children's Surgery department of the C.H.R. of Kénitra (Gharb),

All the children who took part in this epidemiological study underwent a preliminary clinical examination, in order to establish the pathology of each child recruited for the survey. The examination was carried out by paediatricians at El Idrissi Hospital. For reasons of cost and feasibility, most of the PHEs aimed at assessing the prevalence of intestinal parasitism in this study were limited to results obtained with a single parasitological stool examination (PSE). Appropriate anti-parasitic treatment is prescribed to parasitised children by doctors in the event of positive parasitological stool examinations.

II.2.2. Census of children and stool sampling

During the four months of our study, paediatric and child surgery departments were consulted on a daily basis to identify new children admitted to hospital. On each sampling day, one of the parents of the hospitalised child was called. Anyone who turns up is given an identification code, a number and a name. a clean, dry, hard plastic container bearing the child's identification code, in which to collect the child's faeces. They will then be asked to provide their child's surname, first name, age and level of education, as well as other information given in the questionnaire. The data obtained were recorded on specially designed personal cards (APPENDIX 1) with headings for family and socio-economic data, anthropometric data (height and weight), level of hygiene, clinical symptoms, history of intestinal parasitosis, reason for hospitalisation and results of clinical examination.

II.2.3. Applied analysis methods

A. Parasitological stool examination (PSE)

II.2.3.1. Parasitological examination

The stools were coded and examined first at the Medical Analysis Laboratory at El Idrissi Hospital in Kénitra, independently of the results of other samples taken from the patient. The faeces collected in the plastic containers were examined in all cases, less than 24 hours after being passed, first macroscopically to note their consistency and the possible presence of Taenia rings, Ascaris adults or pinworms. They were then examined microscopically, first at low magnification (x100) to detect helminth eggs and larvae, then at medium magnification (x400) to look for vegetative and cystic forms of protozoa (WHO, 2001). A single observation of direct examination or examination with lugol or physiological water is not sufficient to determine the negativity of the sample analysed. Each sample was examined three times. Once the stools have been analysed macroscopically and microscopically, they are preserved in 10% formalin and then analysed in the biology and health laboratory at the Faculty of Science in Kenitra. They were then examined using the enrichment methods recommended by the WHO (2001).

II.2.3.2. Parasitological microscopic examination technique

It looks for helminth eggs and larvae, and vegetative and cystic forms of protozoa. It also indicates the presence of leucocytes, red blood cells, fungal elements, crystals and food residues in the stools. Microscopic examination always involves direct examination and two enrichment techniques.

a. Direct examination

This is the simplest, quickest and least expensive method of observing parasites in faeces. It also shows the mobility of the vegetative forms of protozoa. It involves two preparations, the first with physiological water and the second with lugol.

b. Direct examination with physiological water

Using a shaker, a small particle of faecal matter is removed and spread between a drop of physiological water and a slide covered with a coverslip. The preparation must be thin and examined in its entirety; first at magnification (obj x 10), once the suspect parasitic elements have been identified, they are confirmed at magnification (obj x 40).

c. Direct examination with lugol

It follows the same procedures as above, except that the drop of physiological water is replaced by a drop of lugol.

Lugol :

Iodine = 1 gram

Potassium iodide = 2 grams Distilled water = 100 ml

This solution reveals the internal morphology of the protozoa and the contents of the cysts, making them much easier to identify.

II.2.3.3. Enrichment techniques used

In addition to direct examination, complementary methods are used to diagnose parasites in stools. Enrichment techniques to detect eggs, larvae and cysts and permanent staining techniques to detect vegetative and cystic forms are the most commonly used.

The aim of these enrichment techniques is to concentrate parasite elements that are too rare to be detected by direct examination. The ideal technique for concentrating all parasitic elements does not yet exist (WHO, 2003), which explains the large number of methods proposed.

The two standardised enrichment techniques used in our present study for stool examination are as follows:

a. WILLIS method (WHO, 2001)

It concentrates any parasites present in the stools by flotation. The dilution liquid is a saturated aqueous solution of sodium chloride.Pure crystallised sodium chloride = 250 grams Distilled water = 750 ml

Preparation :

Mix 2 grams of stool in 20 ml of dilution liquid. Fill a haemolysis tube with the mixture obtained until the liquid is flush with the rim and forms a protruding meniscus over which a slide can be passed, taking care not to include any air bubbles. Leave to settle for one hour. The helminth eggs will adhere to the slide. Turn the slide over and cover the area that has been in contact with the liquid with a coverslip, then examine under the microscope. This technique is particularly recommended for the detection of helminth eggs (WHO, 2001).

b. RITCHIE method (WHO, 2001)

A quantity of 2 grams of stool is diluted in 20 ml of 10% formulated water, then sieved through a sieve. 2 ml of the filtrate is collected in a centrifuge tube into which 1 ml of ether is added. Stopper the tube and shake to obtain a homogeneous emulsion. Centrifuge for 2 minutes at 2000 rpm. Four layers are obtained, (1) a slightly yellowish ethereal layer, (2) a thick layer containing debris, (3) a formolised layer and a pellet (4) containing the parasite eggs, cysts and larvae. The tube is emptied quickly. A pipette is used to recover the centrifuged pellet, which is examined between slide and coverslip (Figure 2). This technique is recommended for the detection of protozoa (WHO, 2001).

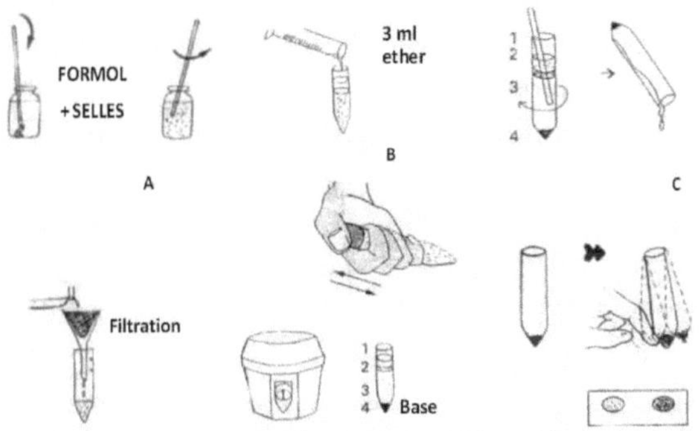

Figure 2: (Diagrams: taken from Manuel des techniques de base pour le laboratoire médical, World Health Organisation, Geneva 1982).

c. The anal Scotch test (WHO, 2001)

Apply the adhesive side of a small tape around the child's anus in the morning before washing and defecation. Stick the tape to a slide and examine under the microscope. This technique is used to look for pinworm eggs (Enterobius vermicularis). In our present study, this technique was only performed on children aged between one and nine years, who had anal pruritus as a warning sign (172 children).

B. Statistical analysis of data

The data were entered and analysed using SPSS software (version 10). The association between potential risk factors and intestinal parasitism was measured using a logistic regression model. The difference was considered significant for p :S 0.05.

III. RESULTS

III.1. Characteristics of the children surveyed

The planned number of tests was 320 samples. However, only 300 were analysed. The remaining 20 were not included in the study for various reasons (defecation impossible, refusal to participate, patient discharge, etc.).

III.1.1. Breakdown of children by department and reason for hospitalisation

The 300 children we examined at the El Idrissi Hospital in Kénitra were distributed as follows: 206 children hospitalised in the paediatrics department, with a percentage of 68.67%, and 94 children hospitalised in the children's surgery department, with a percentage of 31.33% (Table 1). Of the 300 children examined, 42 were hospitalised for acute gastroenteritis (14%), 47 for sickle cell anaemia (15.67%), 47 for bronchopneumonia (15.67%) and 12 for acute appendicitis (4%). The other reasons were : Injuries, Burns, Road accident,(Table 1).

Table 1: Breakdown of children surveyed by department and cause of hospitalisation ParametersN = 300

Service	Workforce	%
Paediatrics	206	68,67
Pediatric Surgery	94	31,33
Total	300	100,00
Cause of hospitalisation	Frequency	%
Sickle cell anaemia	47	15,67
Broncho-pneumonia	47	15,67
Acute gastroenteritis (AEG)	42	14,00
Acute appendicitis (AA)	12	4,00
Hernia	22	7,33
Acute dehydration (AHD)	21	7,00
Meningitis	9	3,00
Food poisoning (AI)	7	2,33
Other reasons (injuries, burns, road accident, etc.)	93	31,00
Total	300	100,00

III.1.2. Breakdown of children surveyed by origin

Table 38 shows that the city of Kénitra had the highest number of children hospitalised: 35.33% of the total sample (106/300). The other children from the region's towns and villages: Sidi Yahia, Sidi Slimane, Souk Larbaa du Gharb and M'nasra had 39, 41, 40 and 38 children respectively (Table 2).

Table 2: Geographical breakdown of children surveyed by origin

Towns/Communes	Workforce	%
Kenitra	106	35,33
Sidi Yahia	40	13,33
Sidi Slimane	41	13,67
Sidi Allal Tazi	18	6,00
Souk Larbaa du Gharb	39	13,00
Belksiri	9	3,00
M'nasra	38	12,67
Mehdia	7	2,33
Sidi Taibi	2	0,67
Total	300	100,00

III.1.3. Breakdown of children by gender, age and environment

The proportion of males is greater than that of females. This difference is statistically significant. Children aged between 1 and 4 accounted for 39.33% of the children surveyed. The other age groups were respectively 21.33%, 21.33% and 18% for the under 1, 10 to 16 and 5 to 9 age groups. The percentage of children surveyed from rural areas is higher than that calculated for children from urban areas (65% in rural areas compared with 35% in urban areas) (Table 3).

Table 3: Breakdown of children surveyed by sex, age and place of residence Parameters N = 300

	Workforce	%
Gender Boys Girls	203 97	67,67 32,33
Age (years)< 1	64	21,33
1 - 4	118	39,33
5 - 9	54	18,00
10 - 16	64	21,33
Environment Urban Rural	105	35
	195	65

III.1.4. Quality of life criteria for children surveyed

Table 4 shows the number of children surveyed by quality of life (persons/household, number of brothers/household, number of rooms/household, type of housing, presence or absence of private toilets, liquid sanitation, household waste collection and drinking water supply).

Table 4: Breakdown of children by quality of life criteria Life criteria

	Workforce	%
People / household :S to 5 > à 5	 158 142	 52,67 47,33
No. of brothers / household :S to 5 > à 5	 275 25	 91,67 8,33
No. of rooms / fireplace :S to 2 > à 2	 224 76	 74,67 25,33
Type of habitat Dur Bidonvilles	 172 128	 57,33 42,67
Private toilet Present Absent	 238 62	 79,33 20,67
Liquid drainage Sewer system Septic tank Open sky	 105 118 77	 35,00 39,33 25,67
lid waste disposal Public collection Collective waste bin Close to the foyer	 82 61 157	 27,33 20,33 52,33
AEP Drinking water network Well Public fountain Other springs (Oued, Sagia, etc.)	 123 107 67 3	 41,00 35,67 22,33 1,00

N = 300

275 cases of children with less than or equal to five brothers per household (91.67%) and 25 cases of children with more than five brothers per household (8.33%). The breakdown by number of rooms per household showed that 74.67% of the children surveyed had more than two bedrooms per household, while the rest had only 25.33%. The majority of the children surveyed (79.33%) had a private toilet in the home, and almost 57.67% of these children lived in solid or semi-solid housing and 42.67% in shanty towns. These figures reflect

the peri-urban or rural nature of the localities of origin of our child population. When we looked at family size, we found that the number of people in the household was the same for both categories (:S to 5 people/household and > 5 people/household).Among the factors in the physical environment that affect health is the supply of drinking water. An analysis of the distribution of the children surveyed shows that 41% are connected to the drinking water network, 35.67% use well water and 22.33% use water from public fountains. According to the availability of private toilets in the home, 20.67% of the children surveyed had a private toilet. We also noted that 25.67% and 39.33% respectively had an open-air or septic tank type of liquid sanitation system, with the remainder connected to the sewage system. In terms of solid waste disposal, 52.33% of the children's families surveyed had household waste disposal close to home, 20.33% had a communal rubbish bin and 27.33% had public rubbish collection. These figures are in line with those of the other parameters used to characterise the peri-urban and rural situation of our population's localities of origin.

III.1.5. Hygiene applications, symptoms and previous parasitosis in the children surveyed

The breakdown by level of hygiene, digestive disorders and previous parasitosis is shown in Table 5. Fifty-eight of the children surveyed (19.33%) had dirty fingernails, indicating poor hand hygiene. Almost 71.33% of the children surveyed used only water to wash their hands before each meal, compared with 2.33% who did not. While26.33% used soap and water before each meal.At the time of the examination, the stoolswere pasty, soft or much more rarely moulded in 64.33%, liquid or semi-liquid in 35.67% of the total sample. We noted the presence of acute diarrhoea (> 3 days) in 23.67% of the children surveyed (71 cases). Only 7.67% (23 cases) and 2.67% (8 cases) had persistent diarrhoea (> 2 weeks) and chronic diarrhoea (> 4 weeks) respectively. The stools were non-diarrhoeal in 198 cases, i.e. 66% of the population surveyed. The clinical picture was dominated by abdominal pain, observed in 34.67% of cases, vomiting (4.33%) and anorexia (4.33%). Anal pruritus was present in 8% of the children surveyed. 48% of the children examined showed no signs of digestive disorders or pain. Bulimia was noted in two hospitalised children. The clinical course was favourable for all the children examined. In terms of a history of parasitosis, 24.67% of the children examined had been infested by digestive parasites, in particular helminth worms (Ascaris, pinworms, etc.).

Table 5: Breakdown of children surveyed by personal hygiene, digestive disorders and previous parasitosis Parameters

	Frequency	%
Nail condition Clean Medium Rooms	 122 120 58	 40,67 40,00 19,33
Hand washing before meals No Yes (water) Yes (water + soap)	 7 214 79	 2,33 71,33 26,33
Digestive disorders		
N = 300		
None Abdominal pain Anorexia Bulimia Vomiting Anal pruritus	144 104 13 2 13 24	48,00 34,67 04,33 00,67 04,33 08,00
Types of diarrhoea Acute diarrhoea > 3 days Persistent diarrhoea > 2 Weeks Chronic diarrhoea > 4 Weeks Non-diarrhoeal stools	 71 23 8 198	 23,67 07,67 02,67 66,00
Aspect of stool Solid - Mouldy Soft Semi-liquid Liquids	 73 120 66 41	 24,33 40,00 22,00 13,67
Previous parasitosis No Yes	 226 74	 75,33 24,67

III.1.6. Education criteria and socio-economic level

Table 6 shows the number of children surveyed by school level. Almost 70% of the children surveyed did not attend school. 4% attended secondary school, mosques and day nurseries. In addition, 23% of those examined had a primary school education.A breakdown by parents' occupation shows that 145 children (48.33%) have a father who is a farmer. This reflects the agricultural vocation of the localities of origin (Souk Larbaa du Gharb, Banlieue de Kénitra, Sidi Slimane, Sidi Yahia and the use of waste water and the Oued Sebou and its tributaries for agriculture. So 38.67% have a father who is an employee (bricklayer, carpenter, labourer, gardener, etc.). The majority of the mothers in our surveyed child population are housewives (292, or 97.33%) and work temporarily in agriculture. Eight mothers (2.67%) were engaged in seasonal work.

Table 6: Breakdown of children surveyed by school level and father's and mother's occupation

Parameters

	Workforce	%
School level Primary School Secondary School Mosque/Childcare Out of school	69 12 12 207	23,00 4,00 4,00 69,00
Father's occupation No Farmer Employee Civil servant Other (Decided, Retired, ...)	12 145 116 15 12	4,00 48,33 38,67 5,00 4,00
Mother's occupation Housewife Employee Civil servant	292 6 2	97,33 2,00 0,67

N = 300

III.2. Overall prevalence of intestinal parasitism among children surveyed

III.2.1. Overall prevalence of intestinal parasitism

Of the 300 children examined, 157 were found to be carrying one or more intestinal parasites, giving an overall prevalence of 52.3% (95% CI: 21.6-39.6). Overall, the results show that one child in two is infected. There was no significant variation in overall prevalence according to sex. Of the 203 boys examined, 114 (56.16%) were infected. Of the 97 girls examined, 43 (44.3%) were parasite carriers. Statistical analysis shows that the difference observed is significant ($p<0.05$).

III.2.2. Study of parasitic associations

Qualitative analysis of polyparasitism revealed 15.2% (24/157) of positive cases in the case of associations with two protozoan species and 3.1% (5/157) of positive cases in the case of associations with three protozoan species (Tables 7 and 8). A only one case of pentaparasitism was noted, i.e. 0.63% of positive cases (Table 9).

Table 7: Number of cases of parasitic associations of two protozoa

Parasitic associations	No. of case
Entamoeba histolytica/dispar + +. Blastocystis hominis	2
Entamoeba histolytica/dispar + +. Entamoeba coli	2
Entamoeba coli + Blastocystis hominis	9
Entamoeba coli + Iodamoeba butsclii	1
Entameoba coli + Giardia intestinalis	5
Giardia intestinalis + Blastocystis hominis	5
Total	24

Table 8: Number of cases of parasitic associations of three protozoa

Parasitic associations	No. of cases
Entamoeba coli + Blastocystis hominis + Iodamoeba butsclii	1
Entamoeba histolytica + Entamoeba coli + Iodamoeba butsclii	1
Entameoba coli + Giardia intestinalis + Blastocystis hominis	2
Entameoba coli + Giardia intestinalis + Iodamoeba butsclii	1
Total	5

Table 9: Number of cases of parasitic associations of five protozoa.

Parasitic associations	No. of case
Entamoeba histolytica/dispar + Entamoeba coli + Giardia intestinalis + Blastocystis hominis + Iodamoeba coli	1
Total	1

Combinations with two helminth species represented 14.6% (23/157) of positive cases (Table 10). 38 children were parasitised by both a protozoan and a helminth, i.e. 24.2% of positive cases (Tables 11 and 12).

Table 10: Number of cases of parasitic associations of two helminths

Parasitic associations	No. of case
Ascaris lumbricoides + Trichuris trichiura	14
Ascaris lumbricoides + Enterobius vermicularis	6
Ascaris lumbricoides + Hymenolepis nana	1
Ascaris lumbricoides + Taenia saginata	1
Enterobius vermicularis + Taenia saginata	1
Total	23

Table 11: Number of cases of mixed associations of two parasites

Parasitic associations	No. of case
Ascaris lumbricoides + Blastocystis hominis	3
Ascaris lumbricoides + Entamoeba coli	6
Ascaris lumbricoides + Giardia intestinalis	1
Trichuris trichuira + Entamoeba coli	1
Enterobius vermicularis + Blastocystis hominis	4
Enterobius vermicularis + Entamoeba histolytica/dispar	2
Enterobius vermicularis + Endolimax nana	5
Enterobius vermicularis + Giardia intestinalis	2
Hymenolepis nana + Blastocystis hominis	1
Taenia saginata + Entamoeba coli	1
Total	26

Table 12: Number of cases of mixed associations of three parasites

Parasitic associations	No. of case
Ascaris lumbricoides + Trichuris trichiura + Entamoeba coli	2
Ascaris lumbricoides + Enterobius vermicularis + Hymenolepis nana	1
Ascaris lumbricoides + Entamoeba coli + Blastocystis hominis	1
Enterobius vermicularis + Blastocystis hominis + Iodamoeba butsclii	1
Enterobius vermicularis + Taenia saginata + Blastocystis hominis	1
Enterobius vermicularis + Entamoeba coli + Blastocystis hominis	4
Enterobius vermicularis + Giardia intestinalis + Iodamoeba butschlii	1
Hymenolepis nana + Entamoeba coli + Blastocystis hominis	1
Total	12

III.2.3. Polyparasitism index (PPI)

This index expresses the number of polyparasitic children in relation to the total number of children examined. Of the 300 children examined, we found 91 polyparasitic children, a rate of 30.3%. Biparasitism accounted for 24.3% (73 cases) of parasitized children, triparasitism for 5.7% (17 cases) and pentaparasitism for 0.3% (1 case) of these positive cases (Table 13).

Table 13: Polyparasitism in the children examined

Number of species of parasites	Workforce	%
2	73	24,3
3	17	5,7
5	1	0,3
PPI	91	30,3%

I.P.P: Index of polyparasitism

III.3. Study of the prevalence of intestinal parasitism

III.3.1. Parasite prevalence by hospital ward

Table 14 shows that the rate of intestinal parasitism was relatively equal in children hospitalised in the paediatric ward (46.12%) and those hospitalised in the children's surgery ward (65.96%). The Chi-2 test revealed a significant difference ($p < 0.05$).

Table 14: Results of parasitological examinations of stools, by hospital ward

	EPS results				Chi-square 2	P	CI (95%)
Service	Children no parasites	%	Children parasites	%			
Paediatrics	111	5,34	95	46,12	10,1	0,000	1,33 - 3,9
Surgery Children	32	34,04	62	65,96			
Total	143	47,67	157	52,33			

III.3.2. Parasite prevalence by reason for hospitalisation

Clinical manifestations leading to hospitalisation were largely dominated by sickle cell anaemia, acute gastroenteritis and acute appendicitis. Out of 47 anaemic children, 33 (70.21%) had parasites. In addition, out of 12 children hospitalised with acute appendicitis, there were 10 cases of intestinal parasitism. Of the 42 children hospitalised with acute gastroenteritis, 13 (30.95%) had parasites (Table 15). Statistical data based on parasitological examination of the stools of hospitalised children therefore suggest a probable relationship between the reason for hospitalisation, particularly anaemia, acute appendicitis (AA) and acute gastroenteritis (AGE), and intestinal parasitism in the children examined ($p < 0.05$).

Table 15: Results of parasitological stool tests, by reason for hospitalisation

	EPS results						
Reason for hospitalisation	Children no parasites	%	Children with parasites	%	Chi-square 2	P	CI (95%)
Sickle cell anaemia	14	29,79	33	70,21			
Broncho-pneumonia	29	61,70	18	38,30			
Acute gastroenteritis (AEG)	29	69,05	13	30,95			
Acute appendicitis (AA)	2	16,67	10	83,33			
Hernia	8	36,36	14	63,64	32,60	0,000	1,10 - 2,40
Acute dehydration (DHA)	16	76,19	5	23,81			
Meningitis	3	33,33	6	66,67			
Food poisoning (IA)	2	28,57	5	71,43			
Other causes (Burners, MVA, Fracture, Fever, ...)	40	43,01	53	56,99			
Total	143	47,67	157	52,33			

III.3.3. Parasite prevalence by children's town of origin

Table 16 shows the number and percentage of children with parasites according to the different localities of origin of the child population surveyed at El Idrissi Hospital. The prevalence of intestinal parasitosis is higher among children from towns or predominantly rural communes (Belksiri, Souk Larbaa du Gharb, Sidi Allal Tazi, Sidi Sliman, Sidi Yahiya and M'nasra) located to the west on the Atlantic coast. They are lower among children living in the town of Kénitra centre and the commune of Mehdia. The distribution of the prevalence of intestinal parasitism according to the origin of the children examined shows a predominance among subjects from communes or localities characterised by a low population size and a less developed infrastructure ($p<0.000$).

Table 16: Results of parasitological examinations of stools by origin of hospitalised children

	EPS results				
Towns/Communes	No. of cases negative	%	No. of cases positive	%	Total
Kenitra	65	61,32	41	38,68	106
Sidi Yahia	21	52,50	19	47,50	40
Sidi Slimane	16	39,02	25	60,98	41
Sidi Allal Tazi	6	33,33	12	66,67	18
Souk Labaa from Gharb	13	33,33	26	66,67	39
Belksiri	2	22,22	7	77,78	9
M'nasra	15	39,47	23	60,53	38
Mehdia	4	57,14	3	42,86	7
Sidi Taibi	1	50,00	1	50,00	2
Total	143	47,67	157	52,33	300

III.3.4. Parasite prevalence by sex

A total of 157 children tested positive for parasites, including 114 boys (56.16%) and 43 girls (44.33%), giving a sex ratio of 2.6. The Chi-2 test revealed a significant difference ($p < 0.05$) (Table 17).

Table 17: Results of parasitological stool tests by sex

Gender	EPS results Children without parasites Children with parasites %				Chi-square 2	P	CI (95%)
Boy	89	43,84	114	56,16	3,7	0,03	0,37 - 1,04
Girl	54	55,67	43	44,33			
Total	143	47,67	157	52,33			

III.3.5. Parasite prevalence by age

The average age of the 300 children surveyed in the paediatric and child surgery departments was 57.7 ± 52.47 months (1 month - 16 years). The age distribution of the prevalence of intestinal parasitism shows a predominance among children aged between 5 and 9 years (Table 18). We also note that children are infested from an early age and even before the age of one (4.69%). The Chi-2 test revealed a highly significant difference ($p < 0.05$).

Table 18: Stool parasitology results by age

Age group	Children without parasites	EPS results %Children with parasites		%	Chi-square 2	P	CI (95%)
< 1 year	61	95,31	3	4,69			
1 - 4 years	44	37,29	74	62,71	23,5	0,000	0,98 - 0,90
5 - 9 years	12	22,22	42	77,78			
10 - 16 years	26	40,63	38	59,38			
Total	143	47,67	157	52,33			

III.3.6. Parasite prevalence by place of residence

Table 19 shows that 31.43% or 33 of the 105 children examined who lived in urban areas were carriers of intestinal parasites. This rate rose to 63.59% in children living in rural areas. The prevalence of intestinal parasitism was significantly higher in rural than in urban areas ($p < 0.05$).

Table 19: Stool parasitology results by place of residence

Environment	Children without parasites	%	Children with parasites	%	Chi-square 2	P	CI (95%)
Urban	72	68,57	33	31,43	28,3	0,000	2,22 - 6,56
Rural	71	36,41	124	63,59			
Total	143	47,67	157	52,33			

III.3.7. Parasite prevalence according to lifestyle criteria

According to the results shown in table 20, there was no difference between the risks of having intestinal parasitism according to the number of people living in the same household ($p > 0.05$). However, there was a significant difference in the risk of intestinal parasitism according to the number of brothers living in the same household ($p < 0.05$). There was also no significant association between intestinal parasitism and the number of rooms in the same household ($p > 0.05$). Living in a solid dwelling would appear to be a protective factor against infestation by intestinal parasites ($p < 0.05$). Table 56 shows the distribution of children surveyed according to the presence or absence of private toilets in their homes. This shows that children living in homes with private toilets were as often parasitized (49.16%) as children living in homes without private toilets (64.52%)($p < 0,05$). Connection to the liquid sewage system appears to protect against infestation by intestinal parasites. There was a significant association between intestinal parasitism and the type of liquid sanitation: septic tanks and open discharge ($p < 0.05$). For solid sanitation, the Chi-2 test revealed a significant difference ($p < 0.05$). In addition, the availability of drinking water and connection to the network appeared to protect against intestinal parasitism ($p < 0.05$).

Table 20: Results of parasitological examinations of stools according to children's life criteria

	Results		EPS		Statistical analysis		
No. of people per household	Children without parasites	%	Children with parasites	%	Chi-square 2	P	CI (95%)
:S to 5	85	53,8	73	46,20			
> à 5	58	40,85	84	59,15	1,6	0,207	0,69 -
Total	143	47,67	157	52,33			1,08
No. of brothers per household	Children without parasites	%	Children parasites	%	Chi-squa	P	IC (95%)
:S to 5	134	48,73	141	51,27			
> à 5	9	36,00	16	64,00	14,3	0,050	
	143	47,67	157	52,33			2,15
No. of rooms / fireplace	Children without parasites	%	Children parasites	%	Chi-squa	P	IC (95%)
:S to 2	113	50,45	111	49,55			
> à 2	30	39,47	46	60,53	7,7	0,360	
Total	143	47,67	157	52,33			1,10
Type of habitat	Children without parasites	%	Children parasites	%	Chi-squa	P	IC (95%)
Hard	95	55,23	77	44,77			
Garbage cans	48	37,50	80	62,50	10,3	0,006	
Total	143	47,67	157	52,33			0,83
Private toilet	Children no	%	Children		Chi-	P	IC
parasites			parasites		2		(95%)
Present	121	50,84	117	49,16			
Absent	22	35,48	40	64,52	4,7	0,030	
Total	143	47,67	157	52,33			3,53
Liquid sanitation	Children no parasites%		Children parasites	%	Chi-squa	P	IC (95%)
Network	69	65,71	36	34,29			
Pit	48	40,68	70	59,32	21,9	0,000	0,45 -
Open sky	26	33,77	51	66,23			1.47
Total	143	47,67	157	52,33			
Solid waste treatment	Children no parasites%		Children parasites	%	Chi-squa	P	IC (95%)
Public collection	56	68,29	26	31,71			
Collective bin	29	47,54	32	52,46	21,2	0,000	0,48 -
Close to the foyer	58	36,94	99	63,06			1.18

Total	143	47,67	157	52,33			
AEP	Children without parasites	%	Children parasites	%	Chi-squa	P	IC (95%)
Drinking water	77	62,60	46	37,40			
Well	46	42,99	61	57,01			
Public fountain	19	28,36	48	71,64	22,2	0,000	0,44 - 1,10
Other sources							
Sagia, ...)							
Total	143	47,67	157	52,33			

III.3.8. Parasite prevalence according to personal hygiene, digestive disorders and previous parasitosis

Table 21 shows that only 122 of the children examined had a correct level of hygiene in terms of the condition of their nails. Of the 122 children, 38 (31.15%) had parasites.Thus, 42 out of 58 children (72.41%) and 77 out of 120 children examined (64.17%) were parasitized and had an average or poor level of hygiene in terms of the cleanliness of their nails. The Chi-2 test revealed a significant difference ($p < 0.05$).22 children out of 79 (27.85%) surveyed who stated that they correctly washed their hands with soap and water before each meal were infected, and 131 children out of 214 (61.21%) who only occasionally took this precaution were infected (washing only with water). While 4 out of 7 children who did not wash their hands with soap and water were infected. The influence of individual hygiene is considerable, as shown by the statistical analysis ($p < 0.05$).Acute diarrhoea was noted in 32 of the parasitized children (45.07%), persistent diarrhoea in 9 cases among 23 children (39.13%) and in 5 parasitized children, the diarrhoea was chronic (62.5 %). Of the 198 children examined with non-diarrhoeal stools, 111 (56.06%) were found to have parasites. There was no difference between intestinal parasitosis and stool consistency (p= $p > 0.05$).

Other functional signs included: 6 cases of vomiting (46.15%), 24 cases of anal pruritus (100%) and 58 cases of diffuse abdominal pain (55.77%) in children with parasites. Of the 144 children with no digestive problems, 62 were found to be infected (43.06%). The difference was statistically significant ($p < 0.05$). 79.73% of the 74 children examined who had a history of intestinal parasitosis were parasite carriers, compared with 43.36% of the 226 children surveyed who did not have a history of intestinal parasitosis.

Table 21: Results of parasitological stool tests according to personal hygiene, digestive disorders and previous parasitosis in the children surveyed

Parameters	EPS results				Statistical analysis		
Nail condition	Children without parasites	%	Children with parasites	%	Chi-squa re 2	P	CI (95%)
Own	84	68,85	38	31,15	38,1	0,000	0,26 - 0,52
Medium	43	35,83	77	64,17			
own							
Room	16	27,59	42	72,41			
Total	143	47,67	157	52,33			
Hand washing before meal	Children without parasites	%	Children	%	Chi-squa	P	IC (95%)
No	3	42,86	4	57,14			
Yes (water)	83	38,79	131	61,21	25,8	0,000	2,01 -
Yes (water and	57	72,15	22	27,85			5,73
Total	143	47,67	157	52,33			
Digestive disorders	Children without parasites	%	Children parasites	%	Chi-squa	P	IC (95%)
No	82	56,94	62	43,06			
Pain	46	44,23	58	55,77			
abdominal							
Anorexia	8	61,54	5	38,46	30,4	0,000	
							2,15
Bulimia	0	0,00	2				
			0				
Vomiting	7	53,85	6	46,15			
Anal pruritus	0	0,00	24	100,0 0			
Total	143	47,67	157	52,33			
Types of diarrhoea	Children no parasites	%	Children parasites	%	Chi-squa	P	IC (95%)
Acute diarrhoea	39	54,93	32	45,07	4,54	0,208	0,45 - 1,58
Days							
Persistent	14	60,87	9	39,13			
> 2 weeks							
Chronic							
> 4 weeks	3	37,5	5	62,5			
No stool	87	43,94	111	56,06			
diarrhoea							
Total	143	47,67	157	52,33			

Stool appearanceChildren not% Children % Chi-square P IC parasitised parasites 2 (95%)							
Solid - Moulded	35	47,95	38	52,05			
Soft	48	40,00	72	60,00			
Semi-liquids	31	46,97	35	53,03	3.68	0.060	
Liquids	29	70,73	12	29,27			1,01
Total	143	47,67	157	52,33			

ParasitesChildren not % Children % Chi-square PCI antecedents parasites parasites 2 (95%)							
No	128	56,64	98	43,36			
Yes	15	20,27	59	79,73	29.5	0.000	
Total	143	47,67	157	52,33			9,59

III.3.9. Prevalence of parasites according to the level and mother's and father's occupation

The father's profession is a factor influencing the occurrence of intestinal parasitism. In fact, we obtained high percentages among the children of farmers (90 parasitised children out of 145, i.e. 62.07%). The children of farmers were more exposed to intestinal parasitosis than their counterparts whose fathers were not farmers. However, the results obtained show a significant difference between children from different occupational categories of the father ($p < 0.05$). Parasite prevalence is also higher among children whose mothers work in agriculture and in the home (53.42%), whereas it is only 16.67% among children whose mothers work outside the home (employed) (Table 22). Statistical analysis shows that the difference observed is not significant ($p < 0.05$).

Table 22: Results of parasitological examinations of children's faeces by level of education and father's and mother's occupation

School level	Children without parasites	EPS results %Children with parasites		%	Chi-2	P	CI (95%)
Primary	30	37,04	51	62,96	17,8	0,570	1,03 - 1,2
College	5	41,67	7	58,33			
Out of school	108	52,17	99	47,83			
Total	143	47,67	157	52,33			
Father's profession	Children no parasites	%	Children parasites	%	Chi-squar	P	CI (95%)
Without	7	58,33	5	41,67	14,6	0,012	0,94 - 1,6
Farmer	55	37,93	90	62,07			
Employee	65	56,03	51	43,97			
Civil servant	11	73,33	4	26,67			
Other (Retired, Deceased)	5	41,67	7	58,33			
Total	143	47,67	157	52,33			
Mother's occupation	Children no parasites	%	Children parasites	%	Chi-squar	P	CI (95%)
Housewife + Farmer	136	46,58	156	53,42			
Employee	5	83,33	1	16,67	5.4	0.067	0.97 - 7,35
Civil servant	2	100,00	0	0,00			
Total	143	47,67	157	52,33			

III.4. Specific prevalence of intestinal parasitism in children surveyed

Of these children surveyed, 73 cases, or 24.3%, were amoeba carriers; 50 children examined were parasitised by Entamoeba coli (Photo 1), i.e. 16.6% of the cases examined, and Entamoeba histolytica/dispar (Photo 2) was found in 12 children, i.e. a prevalence rate of 4%. Iodomoeba butschili (Photo 3) was diagnosed in 10 children, i.e. a prevalence of 3.3%, and there was only one case of intestinal parasitism due to Endolimax nana (Photo 4), i.e. a prevalence of 0.3% of the examinations carried out. (Table 23). In the flagellate group, a single species, Giardia intestinalis (Photo 5), accounted for 7.3% (22 cases). The species Blastocystis hominis also accounted for 15% (45 cases) (Photo 6). In the group of helminths, pinworms come first, with Enterobius vermicularis (Photo 7) alone accounting for 18.3% of parasitic worms found (55 cases) and 35% of positive tests. The other two nematodes found in the stools of the children examined were Ascaris lumbricoides (Photo 8) and Trichuris trichuira (Photo 9), with prevalence rates of 15.3% (46 cases) and 5.7% (17 cases) respectively. As for cestodes, the main ones were Hymenolepis nana (Photo 10) and Taenia saginata (Photo 11), with very low prevalence rates of 2% and 1.3% respectively (Table 12).

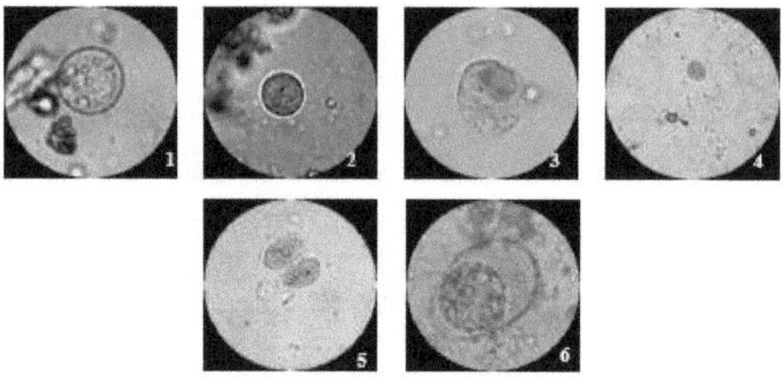

Protozoan cysts :

Entamoeba coli (Photo 13), Entamoeba histolytica/dispar (Photo 14); Iodomoeba butschili (Photo 15), Endolimax nana (Photo 16), Giardia intestinalis (Photo 17), Blastocystis hominis (Photo 18).

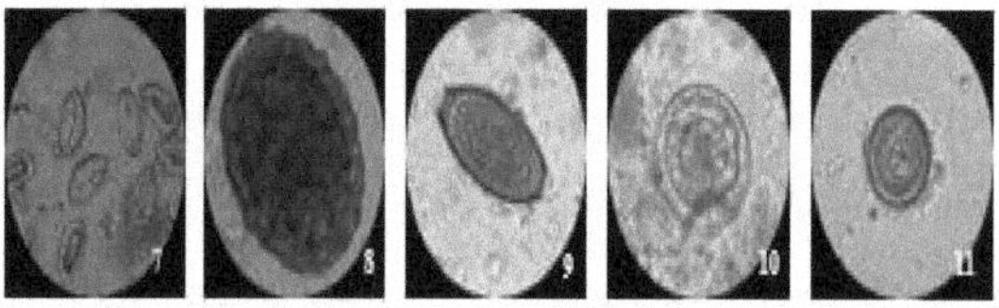

Helminth eggs: Enterobius vermicularis (Photo 19), Ascaris lumbricoides (Photo 20), Trichuris trichuira (Photo 21), Hymenolepis nana (Photo 22) and Taenia saginata (Photo 23)

Table 23: Total and specific prevalence of different parasitic species

PARASITES	Total	Prevalence (%)
HELMINTHES		
Nematodes		
Enterobius vermicularis	55	18,33
Ascaris lumbricoides	46	15,33
Trichuris trichura	17	5,70
Total Nematodes	118	39,37
Cestodes		
Hymenolepis nana	6	2,00
Taenia saginata	4	1,33
Total Cestodes	10	3,33
Total helminths	128	42,70
PROTOZOIRES		
Amoebae		
Entamoeba coli	50	16,67
Entamoeba histolytica/dispar	12	4,00
Iodomoeba butschili	10	3,33
Endolimax nana	1	0,33
Total Amoebae	73	24,33
Flagellated		
Giardia intestinalis	22	7,33
Other protozoa		
Blastocystis hominis	45	15,00
Total Protozoa	140	46,67
TOTAL	268	89,37

Table 23 shows the species distribution of the 157 positive samples. It shows that protozoa accounted for 46.6% of the parasites found, and that their presence in the stools examined was greater than that of helminths. Entamoeba coli appears to be the most common parasite, with 16.6% of healthy carriers (41 cases in boys and 9 cases in girls), followed by Blactocystis hominis with 15% (36 cases in boys and 9 cases in girls). A single Endolimax nana cyst was detected in a boy. Among helminths, Enterobius vermicularis accounted for 35.03% of all positive tests, i.e. a prevalence of 18.33%, followed by Ascaris lumbricoides with a prevalence of 15.33%. The majority of parasites observed were found in males. Sex seems to play a role in their distribution (Table 24). The Chi-2 test revealed a significant difference ($p<0.05$). A male predominance was observed for Entamoeba coli, Blactocystis hominis and Enterobius vermicularis. However, for Hymenolepis nana and Taenia saginata, girls were affected in the same way as boys.

Table 24: Total and percentage of different parasitic species according to sex

PARASITES	Boys	Girls	Total	% compared to exams positive	Prevalence (%)
HELMINTHES					
Nematodes					
Enterobius vermicularis	35	20	55	35	18,3
Ascaris lumbricoides	33	14	47	29,9	15,7
Trichuris Trichura	10	6	16	10,2	5,3
Total Nematodes	78	40	118	75,2	39,3
Cestodes					
Hymenolepis nana	3	3	6	3,8	2,0
Taenia saginata	3	1	4	2,5	1,3
Total Cestodes	6	4	10	6,4	3,3
Total helminths	84	44	128	81,5	42,7
PROTOZOIRES					
AMIBES					
Entamoeba coli	41	9	50	31,8	16,7
Entamoeba histolytica/dispar	9	3	12	7,6	4,0
Iodomoeba butschili	9	1	10	6,4	3,3
Endolimax nana	1	0	1	0,6	0,3
Total Amoebae	60	13	73	46,5	24,3
FLAGS					
Giardia intestinalis	18	4	22	14	7,3
OTHER PROTOZOA					
Blastocystis hominis	36	9	45	28,7	15,0
Total Protozoa	114	26	140	89,2	46,7

Taking into account the overall percentage of positive examinations, we note that the children examined are infested from an early age and even before the age of one (3.18% of positive examinations, i.e. a prevalence of 1.6%). Parasitism peaks between the ages of 1 and 9, the age when children are sent to nurseries and primary schools, and when overcrowding and playing in groups and with dirty soil are likely to encourage parasite infestation (Table 25).

Table 25: Specific prevalence of intestinal parasitism by age group

PARASITES	Total	< 1 year	1 - 4 years	5 - 9 years	10 - 16 years
Helminths					
Nematodes					
Enterobius vermicularis	55	0	30	15	10
Ascaris lumbricoides	46	1	16	14	15
Trichuris Trichura	17	1	5	4	7
Total Nematodes	118	2	51	33	32
Cestodes					
Hymenolepis nana	6	0	1	3	2
Taenia saginata	4	0	1	2	1
Total Cestodes	10	0	2	5	3
Total helminths	128	2	53	38	35
PROTOZOIRES					
Amoebae					
Entamoeba coli	50	1	22	18	9
Entamoeba histolytica/dispar	12	1	5	3	3
Iodomoeba butschili	10	0	2	7	1
Endolimax nana	1	0	0	0	1
Total Amoebae	73	2	29	28	14
Flagellated					
Giardia intestinalis	22	0	14	5	3
OTHER PROTOZOA					
Blastocystis hominis	45	1	21	15	8
Total Protozoa	140	3	64	48	25

IV. DISCUSSION

Intestinal parasitoses seem to be attracting less and less interest these days, as available resources are justifiably being diverted towards other priorities such as malaria, bilharzia and leishmaniasis. These infestations should not be neglected and relegated to second place, given the public health problems they pose for our developing countries.Research is still needed to better define the problem in all regions in order to eradicate this disease forever. Our study, which responds to these various concerns, aims to establish the prevalence of intestinal parasitism in children hospitalised at the Gharb regional hospital (Kenitra) for a pathology (x) and to identify the most frequent species.Intestinal parasitism is a classic phenomenon in children and plays an important role in childhood pathology as a whole, particularly in Third World countries. Because of the all too often precarious climatic and hygienic conditions, intestinal parasitosis plays an equally important role in parasitic diseases as a whole, especially in children (SHUBAIR et al., 2000).Coproparasitological analyses undertaken to assess the frequency of these intestinal parasites are intermittent, rare and difficult to compare due to the wide variability in analysis techniques. In addition, the prevalence of different parasite species varies from one region to another.We wanted to know the specific distribution among children in the Gharb region (Kénitra) and we report the first study concerning 300 parasitological examinations of the faeces of children hospitalised in the paediatric and child surgery departments of the El Idrissi Hospital in Kénitra (Gharb region). This survey provided information on the prevalence of various intestinal parasitoses, as well as the distribution of these parasitoses according to several possible risk factors.The sampling method chosen allowed for a fairly even distribution, resulting in a sample that was fairly representative of the child population in the different areas of the Gharb region (Kénitra).The results showed that 52.3% of the children examined harboured one or more intestinal parasites, i.e. one child in two. This overall prevalence rate is comparable to those of TLIGUI and AGOUMI (2006) in Tiflet (Morocco) in 2004 and NDIR et al. in 2001 at the Centre de Santé Roi Baudouin in Guediawaye (Senegal), which were 57.1% and 56.6% respectively. This is higher than the 25.09% found by AYADI et al (1991) at the Sfax University Hospital (Tunisia) in 3025 children aged 0-10 years, i.e. one child in four, and the 16.32% and 31.51% found by HAJFANI (1976) in infants and children respectively at the Guediawaye Health Centre (Senegal).Ibn Sina University Hospital (Rabat-Morocco). In Walo, Senegal, FAYE et al (1998) found a prevalence rate of 36.5% (5), while in Algérois, Algeria, BACHTA et al (1990) found 37.04% out of 3,364 examinations of children aged 6 to 15, i.e. one child in three, and ADOU-BRYN et al (2001) found 38.9% out of 357 examinations of children aged 5 to 15 in Toumodi, Côte d'Ivoire.Of 6,439 tests, 2009 (31.2%) were positive (GBADOE and KOFFI, 2005) in children under five years of age undergoing consultation and/or hospitalisation in the paediatric wards of the Tokoin Regional Hospital (Togo). These differences can be attributed to the sampling and the study period, but are probably also linked to geographical differences, which are themselves related to different modes of transmission. A distinction is made between "parasitism", which concerns asymptomatic carriers, and "parasitosis", which concerns subjects who are ill because of the parasite.The results of our study show that all towns in the Gharb region of origin of the hospitalised children are affected, although there are disparities in terms of intestinal parasitism. The prevalence of this parasitism was significantly greater in rural than in urban areas ($p < 0.05$).The state of the environment and the behaviour of these populations partly explain this epidemiological situation. There are many places where surface water and groundwater are invaded by industrial, agricultural and household waste. According to the World Water Commission for the 21ème century, more than half of the world's major rivers are so impoverished and polluted that they are endangering human health and poisoning the surrounding ecosystems (KING, 1990).In fact, the populations of the rural villages from which the hospitalised children originate, located on the Gharb region's main watercourse: oued Sebou and its tributaries (oued Beht at Sidi Slimane, oued R'dom at Souk Larbaa, oued Tiflet at Sidi Yahia, oued Sebou at Belksiri, Sidi Allal Tazi, M'nasra and Kénitra), reuse the water from these oueds for agricultural purposes and are highly exposed to pollution from the region's urban discharges. Given the proximity of these towns and the similarity of their social and health environments, we note some curious disparities in the profile of intestinal parasitism. This observation has also been made by other authors (JOMBO et al., 2007, DIANO et al., 2004 and DIOUF et al., 2000).Epidemiological studies in Morocco show that high prevalence of intestinal parasitosis has been found in children exposed to wastewater. These prevalences are significantly higher in the population that uses raw wastewater in agriculture (EL KETTANI and AZZOUZI, 2006; EL KETTANI et al., 2006; LAMGHARI MOUBARRAD, 2005; AIT MELLOUL and AZZOUZI, 2006).al, 2002; ABOUELOUAFA, 2002; HABBARI, 1992-2000; BOUHOUM and SCHWARTZBROD, 1998; BELGHYTI et al, 1994 and BOUHOUM, 1996).Although the figures given for prevalence by town of origin should be treated with caution, they are nonetheless indicative and suggest a link between the occurrence of intestinal parasitism in hospitalised children and socio-economic conditions.

Our study shows that intestinal parasitism affects boys more than girls ($p < 0.05$) and that parasite carriage is highly significantly related to age ($p < 0.05$). These results are consistent with comparable studies (ADOU- BRYN et al., 2001 and MENAN et al., 1997). The authors attribute this difference to the fact that young boys play in gangs around rural settlements, where villagers choose to defecate, while girls stay at home to help with housework (DEVELOUX et al., 1986). On the other hand, in Beni-Mellal, HABBARI et al (2000) reported that, depending on the age and sex of the 1,343 children examined, the differences were not significant.Our results also show that this parasitism is linked to personal hygiene and the source of the water supply. A study carried out in the ASIR region in south-eastern Saudi Arabia revealed that the use of non-disinfected water for drinking was the only risk factor responsible for the high prevalence of intestinal parasitosis (OMAR et al., 1995). This is further supported by studies carried out directly on water sources, which revealed that untreated water was a major vector of intestinal protozoa (FRASER and COOKE, 1991 and HASLAY and LECLERC, 1993). Firstly, children who did not follow the rules of personal hygiene were more infested than those who did ($p < 0.05$), and secondly, those who obtained their water from wells, public fountains or springs (wadis) were more infested than those who had water supplied to their homes ($p < 0.05$). On the other hand, EL KETTANI et al (2006) showed that analysis of the three hygiene factors studied in Settat (Morocco): the source of drinking water, the presence of private toilets in the home and hand washing before eating did not show any significant effect on the prevalence of giardiasis. In children hospitalised at the Guediawaye health centre (Senegal), the difference in parasite prevalence according to sex and the presence or absence of running water was not significant (NDIR et al., 2002). The low rate of parasitized children (34.2%) was observed in towns with a liquid sewage system and a private toilet in the home ($p < 0.05$). Children in towns with no sewage disposal system or where the system is represented by pits and open drains appear to be more infested (59.3% and 66.2% respectively). Statistical analysis of these results shows that there is a significant difference ($p < 0.05$). These results are in agreement with those found by LAAMRANI et al (1999) in the provinces of Beni-Mellal, Taounate and Tiznit (Morocco). In Abidjan (Côte d'Ivoire), the relatively high prevalence of intestinal parasitosis is indicative of a defective sanitation system (MENAN et al., 1997). According to BUCHY (2003), the demographic explosion and the trend towards urbanisation of disadvantaged populations are leading to an increase in the number of shantytowns, raising fears of a further rise in cases of intestinal parasitosis in the future.In our study, living in a solid dwelling was a protective factor against infestation by intestinal parasites ($p < 0.05$). There was no difference in the risk of intestinal parasitism according to the number o f people living in the same household ($p > 0.05$). Surveys of intestinal parasitism in children in Sidi Yahia du Gharb, M'nasra and Marrakech have shown that the rate of parasite infestation increases with the number of people per household and therefore per dwelling (MELYANI, 1983, AMAHMID, 2002). This made it possible to deduce the impact of promiscuity on intestinal parasitism. However, there was a significant difference between the risk of intestinal parasitism according to the number of brothers living under the same roof ($p < 0.05$). For many authors, the socio-economic and hygiene levels are important, even determining, factors in the transmission of intestinal parasitosis, particularly in tropical areas (MENAN et al., 1997; DIOUF et al., 2000; ADOU-BRYN et al., 2001 and BUCHY, 2003). Children examined with a history of intestinal parasitosis are more infested than those without a history of intestinal parasitosis ($p < 0.05$).Contrary to expectations, intestinal parasitism was not related to stool consistency ($p>0.05$). This unexpected result could be due to the association of intestinal parasitism with other pathogenic micro-organisms (bacteria, viruses, fungi, etc.) and the functioning of the digestive tract of the children examined. In terms of the socio-economic and educational background of our child population, there was no significant relationship between parasitism and the level of education of the children examined ($p=0.570$). On the other hand, parasitism was linked to the occupation of the children's parents ($p=0.012$; $p=0.067$). According to LARIVIERE (1987), occupation is a factor favouring the occurrence of intestinal helminthiasis. A retrospective study carried out at the Institut Pasteur in Cocody (Côte d'Ivoire) showed that the prevalence of intestinal parasitism is very high among farmers and breeders, as well as among housewives and domestic servants (MENAN et al., 1997). However, PITISUTTITHUM et al (1990) in Thailand showed that the level of education and social status did not significantly influence infestation by intestinal parasites Our own results show that the most parasitised children have a low socio-economic level and belong to a large family ($p=0.207$). In multivariate analysis, the only factor that could be considered to be significantly and independently linked to intestinal parasitism was the parents' occupation. At the Roi Baudouin health centre in Guediawaye (Senegal), NDIR et al (2002) reported that children hospitalised with a low standard of living had a higher infestation index. In the children examined, we found the classic factors favouring intestinal parasitism, i.e. a low socio-economic level and a large number of siblings. In fact, intestinal parasitism becomes more significant as one moves down the scale of economic status, i.e. parents' professions, and as the number of brothers with whom one lives increases. The trend is less clear-cut for the level of education of the children examined, which might have been expected to correlate inversely with parasitism ($p > 0,05$).Clinically, our children were hospitalised mainly for sickle cell anaemia, acute gastroenteritis and abdominal pain. The most common general and functional clinical signs noted during the course of various intestinal parasitoses were abdominal pain and anal pruritus (100% and 55.7% respectively). Children examined with abdominal pain and anal pruritus were significantly more likely to have parasites. These results are confirmed by those of studies carried out in the internal medicine and infectious diseases departments at Point "G" Hospital in Bamako, Mali (MINTA et al., 2007). Generalized pruritus was associated in 54.28% of cases, abdominal pain in 25.71% of cases. In 2000, a study involving various hospital facilities in Bamako (Mali) showed that isolated diarrhoea was the most frequent reason for hospitalisation, accounting for 13.11% (CISSE et al. 2002). The presence of diarrhoeal syndrome was associated with the presence of parasites ($p=0.0001$) in this study (CISSE et al. 2002). Diarrhoeal diseases represent a real public health problem throughout the world, where they are responsible for 10 to 15 million deaths a year (GENTILINI and DUFLO, 1986). In addition, environmental conditions and cultural habits have a considerable influence on diarrhoeal diseases in Africa (EHOLIE et al., 1999).A prospective study of 150 cases of diarrhoea hospitalised at the Man Regional Hospital (Côte d'Ivoire) showed that chronic diarrhoea was found in the same proportions as acute diarrhoea (EHOLIE et al., 1999). Most patients, and some doctors, believe that digestive parasites can cause diarrhoea. However, diarrhoea is not a usual or even frequent symptom of parasitic infestations. In tropical areas, almost all children are carriers of digestive parasites, but these cause only around 5% of acute diarrhoea and a few chronic diarrhoeas if they are in a good nutritional state (GENDREL, 2003). Parasitic diarrhoea, dominated by helminthiasis, remains a concern at the Man CHR. They are linked to poor hygiene conditions and inadequate health education, a consequence of a low socio-economic level (EHOLIE et al., 1999). In a study carried out at the children's hospital in Libreville (Gabon) on children hospitalised with acute diarrhoea of parasitic origin, the results showed that helminths rarely cause diarrhoea, with the exception of anguillula (Strongyloides stercoralis) in cases of malnutrition, immune deficiency or very heavy infestation. Protozoa (Entamoeba histolytica, Giardia lamblia and Cryptosporidium sp) cause acute diarrhoea much more frequently (GILLES, 1995 and GENDREL, 1990). Furthermore, a study of digestive parasites in children outside tropical hospitals (Togo) showed that the onset of diarrhoea does not seem to depend on the presence of digestive parasites (APLOGAN et al., 1990).Acute gastroenteritis is a widespread attack on the digestive tract, leading to a gastrointestinal symptomatology that is not very specific but is fairly classic.These include diarrhoea, nausea, vomiting and abdominal pain (ARCHAMBAULT et al., 1997; BIGARD, 2001; DEBONNE and REY, 1998; RAMBAUD, 2000). However, the This symptomatology is not always present, and it is not unusual for infection with enteropathogenic diseases to present with even less specific symptoms: fever, weight loss, general malaise, loss of appetite (ARCHAMBAULT et al., 1997 ; BIGARD, 2001 ; DEBONNE et REY, 1998 ; RAMBAUD, 2000), or asymptomatic (MOE, 2001). The duration of the disease is highly variable, generally depending on the aetiological agent, but for most enteric bacterial pathogens varies from 4 days to 2 weeks (BIGARD, 2001; RAMBAUD, 2000). Moreover, acute gastroenteritis (AEG) is the leading cause of death in children in developing countries (GUERRANT and BOBAK, 1991). The responsibility of parasites as agents of AEG varies greatly depending on the populations studied. In temperate countries, the two parasites most often implicated are Giardia intestinalis and Cryptosporidium parvum. The first is most often found in children's communities such as crèches and the second is linked to water contamination (CEVENINI et al., 1985; BRETAGNE et al., 1990). No parasites were found in children under 2 years of age hospitalised for AEG at the Pointe-à-Pitre/Abymes University Hospital, Guadaloupe, during this study (COUROUBLE et al., 2000). The absence of parasites in infantile GEA in our study can be explained by the age of the children and the low yield of enrichment techniques on diluted GEA stools and the limitation to results obtained with a single parasitological stool examination (PSE).It should be noted that the presence of associations with 2, 3 or 5 parasites (67.3%) shows the very low level of food and faecal hygiene, as well as the unfavourable living conditions of these children. The predominance of protozoa is explained by the fact that the parasites concerned often have identical modes of transmission.The majority of protozoa identified are non-pathogenic. They are a sign of poor hygiene, which means that those who carry them are at risk.The results of this prospective epidemiological study, based on hospital recruitment, show that the indices chosen to assess the living conditions of these children hospitalised at the Gharb regional hospital (Kénitra) are certainly not an exact reflection of the level of hygiene, but they do clearly show that the level of hygiene of the environment, housing and drinking water remains fairly low among children mainly from rural areas. Around 65% of the children surveyed are not connected to the sewage system, and almost 60% have no drinking water from the mains. The prevalence of intestinal parasitism is very high among rural children. This is linked to the relatively low level of hygiene in rural areas. Although these results are based on samples taken from children in a hospital environment, the prevalence of intestinal parasitism can be considered high in the Gharb region (Kénitra). This prevalence is slightly lower in urban areas, where hygiene conditions are relatively better. Regular screening followed by appropriate treatment combined with environmental sanitation and hygiene will help to prevent these intestinal parasites. The gender-specific prevalence of intestinal parasitism was 56.1% for boys and 44.3% for girls. The Chi-2 test revealed a significant difference ($p<0.05$). We therefore concluded that boys were more infested than girls. This observation was made in 1995 by AGBERE et al (1995) in children hospitalised in a regional hospital in Togo. The prevalence rates of each parasite are different in the two sexes. However, the Chi-2 test only showed significant differences for Trichuris trichiura, Ascaris lumbricoides and Enterobius vermicularis. In each of the 3 cases, boys were significantly more parasitized than girls. It should be noted that Trichuris trichiura and Ascaris lumbricoides have an identical mode of contamination, whereas Enterobius vermicularis is transmitted directly by dirty hands (LAMGHARI MOUBARRAD, 2005). The difference in prevalence between the two sexes could be explained by the fact that, generally speaking, traditional education in Morocco gives girls and boys different activities; girls generally stay at home on their days off to help their mothers with the housework, and are therefore given responsibilities at an earlier age than boys; as a result, they are more aware of hygiene. Boys enjoy greater freedom, and play games and go on outings where the risk of parasitic contamination could be higher (swimming in wadis or polluted rivers, eating food of dubious quality outside the home, etc.). A study carried out in 1990 at the Centre Hospitalo-Universitaire de Sfax (Tunisia) showed that for Enterobius vermicularis and Hymenolepis nana, girls are more often affected than boys (AYADI et al., 1991). As for protozoa, the two non-pathogenic species Entamoeba coli and Blastocystis hominis are very common in males compared with females. These two protozoa are evidence of poor hygiene, which means that those who carry them are children at risk (LAAMRANI et al., 1999). A study carried out on children hospitalised at Rabat University Hospital in 2001 compared Blastocystis hominis parasitism with the parasites most commonly found in the faeces of this population of children, showing that B. hominis was the most commonly found parasite, with a rate of 46.59%, compared with rates of 37.25% for Entamoeba histolytica, 24.55% for Giardia intestinalis and 17.89% for Entamoeba hartmani (TLIGUI et al, 2002). Prevalence decreases with age. Taking into account the overall percentage of positive cases, we can see that children are infested from an early age and even before the age of one. Parasitism peaks between the ages of 1 and 9, when children are sent to crèches, mosques, nursery schools and primary schools, where overcrowding and playing in groups and on the ground are likely to encourage contamination. This age coincides perfectly with the start of school age. Similar results were reported by AYADI et al (1991) at the University Hospital Centre in Sfax, where children aged 4 to 6 years were more parasitized. On the other hand, in the town of Beni-Mellal, the prevalence of amoebiasis, ascariasis, trichocephalosis, pinworm, hymenolepiasis and taeniasis does not vary according to the age of the children (HABBARI, 1996). Our results confirm the hypothesis that older children are more attentive to sources of pollution and infestation. At Togo's regional hospital, digestive parasitism in hospitalised children was observed from the neonatal period (AGBERE et al., 1995).The parasites found are mainly protozoa. This predominance has also been found in other regions of Morocco (CHOUKRI et al., 2007, KHALLOUFI et al., 2000 ; TLIGUI et al., 2002, HABBARI, 1996, LAAMRANI et al., 1999) and elsewhere (AYADI et al., 1991 and DIOUF et al., 2000). In the group of flagellates, only the species Giardia intestinalis was reported (nearly 14% of all positive cases, i.e. a prevalence of 7.3%). In the group of amoebae, the non-pathogenic Entamoeba coli was the most frequently found (almost 31.8% of all positive cases, i.e. a prevalence of 16.7%). Entamoeba histolytica/dispar was never observed in its haematophagous form, but its cystic form was isolated in 7.6% of positive examinations, which is a significant rate and contrasts with the Tunis study, where this amoeba was practically absent from coprological examinations. in children (BEN SAID, 1981). However, in the Gabès region, 35.80% of parasitised children harbour Entamoeba histolytica (KENNOU, 1979). The protozoa that are usually commensal and whose possible pathogenic role is not always established are rare in our infant population. Entamoeba coli remains the most isolated non-pathogenic protozoan with a prevalence of 16.7%, followed by Iodomoeba butschili and Endolimax nana with prevalence of 3.3% and 0.3% respectively. AYADI et al (1991) reported that within the group of amoebae, the non-pathogenic Endolimax nana was the most frequently found in the faeces of hospitalised children. TLIGUI et al (2006) found E. coli and Endolimax nana with a prevalence of 10.5% and 5.8% respectively. Amoebiasis and giardiasis are common faecal-oral parasites in hot, humid countries. Cysts Entamoeba histolytica and Giardia intestinalis are essential elements in their evolutionary cycles, and transmission is either direct between humans or, more often, indirect via contaminated food or water (BOUREE, 2000, FAYE et al., 1998). In rural schoolchildren in the Man region of Côte d'Ivoire, the predominant intestinal protozoan species were Endolimax nanus (83.8%) and E. coli (74.7%).The prevalence of G. intestinalis and E. histolytica/Entamoeba dispar was 17.5% and 11.3% respectively (OUATTARA et al., 2008). Of the

300 samples examined, 54 were positive for Blastocystis hominis, i.e. 15% (28.7% of positive cases). Nevertheless, our prevalence remains relatively lower than those found by TLIGUI et al (2002) and TLIGUI and AGOUMI (2006), respectively 28.9% (539/4000) and 22.3% (38/170) in children from two different towns (Rabat and Tiflet). On the contrary, our prevalence is higher than that reported by AYADI et al (1991) in Tunis.Analysis of the prevalence of each helminth parasite shows that Enterobius vermicularis (18.3%) is the most common helminthiasis, with a prevalence of 18.3%. It is also the most common in boys aged 1 to 4, with a rate of 11.6%. In a study carried out on children in the paediatric clinic at the Sfax University Hospital, 25% of children systematically sampled had oxyurosis (AYADI et al., 1986). This is a cosmopolitan digestive parasitosis, but the frequency and intensity of parasitism are higher in developing countries with low levels of hygiene. The fact that it is carried in the mouth by fingers soiled by scratching the anal region explains why it is highly contagious in children and, by the same token, contaminates families and schools (KHALLOUFI et al., 2000). The prevalence of 15.7% attributed to ascariasis is similar to those found in Abidjan, Côte d'Ivoire (16.5%) by MENAN et al. (1997) and in El Jadida (18.1%) by LAMGHARI MOUBARRAD (2005). High frequencies have been reported in other regions of Morocco (BOUHOUM and SCHWARTZBROD, 1998; HABARRI et al., 2000 and LAAMRANI et al., 1999). while no case of ascariasis was detected in a crèche in Rabat (KHALLOUFI et al., 2000).In Mexico, CIFUENTES et al (2000) found that children whose families used waste water for crop irrigation were five times more infested with A. lumbricoides than control households. In inter-tropical Africa, the prevalence of this parasite varies. It appears to be most prevalent in humid, rainy forest regions, where eggs are more easily disseminated. The prevalence of trichocephalosis (5.3%) is similar to that reported by MENAN et al (1997) in children hospitalised in Abidjan (5.3%) and in the same child population in Toumodi, Côte d'Ivoire (9.5%) (ADOU-BRYN, 2001). However, no cases of trichocephalosis have been detected in children hospitalised at Rabat University Hospital (ALAOUI et al., 1997) or in children attending the paediatric clinic at Sfax University Hospital (AYADI et al., 1991). In Guatemala: two epidemiological studies by WATKINS et al (1996a-b): one concerned 246 children living in rural areas: 91% were parasitized by roundworms and 82% by whipworms; the other concerned 228 children living in towns; the percentages for the same two worms were identical to those found in the previous study. In Kenya, among 55 children in a primary school, 84% had whipworms and 29% had roundworms (ADAMS, 1994). In South Africa, a study (VERMEULEN et al., 1997) carried out in the Northern Province, among

303 children aged between 4 and 6, found only 5.9% ascaris carriers, while the number of whipworm carriers was insignificant. In Madagascar, in a rainforest region in the south-east of the island, 1,292 children under the age of 11 were studied (KIGHTLINGER et al., 1995). Ascaris was present in 78% of them, and 38% were carriers of whipworms. At the age of 10, the prevalence of ascariasis was 100%.As for cestodoses, Taenia saginata was isolated only 4 times, and given its mode of contamination by ingestion of undercooked meat containing cysticerci larvae, its low frequency in children is understandable. Similarly, Hymenolepis nana accounts for only 3.8% of positive cases, compared with 2.3% among schoolchildren in Tiflet (TLIGUI and AGOUMI, 2006).In Togo, AGBERE et al (1995) reported that this parasitosis affects 3.7% of the child population. Adult cestodes do not therefore appear to be a major parasite in children. Our results are similar to those found by BEN SAID et al (1981) in the Tunis study. In Morocco, previous studies have shown that this parasitosis is more common among children from areas where raw wastewater is reused for agricultural purposes (BELGHYTI et al., 1994; BOUHOUM and SCHWARTZBROD, 1998; HABARRI, 1999; AMAHMID, 2004; LAMGHARI MOUBARRAD, 2005).The predominance of protozoa is explained by the fact that the parasites concerned often have identical modes of transmission. The majority of protozoa identified are non-pathogenic. They are a sign of poor hygiene, which means that those who carry them are at risk. In fact, as a result of a probably identical mode of contamination, these subjects have a greater probability of harbouring, in addition to non pathogenic parasites, either simultaneously or subsequently. A particular feature of protozoan parasites is their ability to form cysts, ensuring their survival in environmental conditions unfavourable to their development (LAAMRANI et al., 1999). The ecology of these organisms in water distribution systems needs to be studied in greater depth, but it seems that filtration and chlorination are suitable treatment methods. It should be noted that the presence of parasitic association shows the very low level of sanitary, food and faecal hygiene, as well as the unfavourable living conditions of these polyparasitic children. Certain intestinal parasitoses are treated on the basis of suggestive clinical signs without recourse to the laboratory, an attitude which runs the risk of ignoring polyparasitism (LAHLOU, 1983).The fight against these intestinal parasites involves education and economic development, but the current state of the economies of most developing countries means that their control will be postponed for the distant future, especially as this fight is not considered a priority (ULUKANLIGIL et al. 2001; HAMADA et al., 2003).

V. CONCLUSION

Our epidemiological survey revealed that the prevalence of intestinal parasitism among children in hospitals in Kénitra in 2007 was 52.7%. Parasitological examination of stools does not give a true picture of the extent of parasite infestation, as examination of a single stool sample detects the parasite in only 50 to 70% of cases. Nevertheless, our prevalence remains relatively representative, as it concerns a targeted population, since parasitological analysis of stools is carried out systematically in all hospitalised children, and not only in children presenting a digestive warning sign. Furthermore, this rate of parasitism may be considered to be underestimated, since it concerns a population of children mainly from the Gharb region, where hygiene conditions and eating habits clearly differ from those in other regions of Morocco. However, it has to be said that the rate of 52.7% is still high, and should encourage us to redouble our efforts in the fight against faecal peril and intestinal parasitosis. In view of the results obtained, we can deduce that children with a low socio-economic level are the most exposed to the risk of parasite infestation. The synergy of all these socio-economic factors favours an increase in the frequency of infestation by parasites. We also deduce that children living in poor hygiene conditions are infested. The quality of hygienic conditions is a factor influencing the health and hygiene status of the child population infested by intestinal parasites. In our study, the deterioration of hygienic conditions and the environment has remarkably increased the rate of intestinal parasitism in children in the Gharb region (Kénitra).These results constitute a point of agreement between the various studies carried out in Morocco, in particular LAAMRANI et al, (1999). At the end of this copro-parasitological analysis, helminth eggs and protozoan cysts were found in the faeces of the infant population of the Kénitra region (Gharb), and at the same time, we were able to provide details of the species likely to be found in the domestic wastewater of Kénitra. Given the importance of other parasitic infestations, further work should also be carried out on cryptosporidiosis and microsporidiosis, which were not reported in this study.

VI. REFERENCES BIBLIOGRAPHIQUES

- **Abouelouafa M. 2002.** "Traitement et réutilisation des eaux usées en agriculture au Maroc Oriental (Oujda): Etude physico-chimique, agronomique et sanitaire" Doctoral thesis. Fac. Sci. Oujda. 141p.
- **Adams EJ, Stephenson LS, Latham MC & Kinoti SN. 1994.** Physical activity and growth of kenyan school children with hookworm, Trichuris trichiura and Ascaris lumbricoides infections are improved with albendazole. J Nutrition, 124, 1199 - 1206.
- **Adou-Bryn D., Kouassi M., Brou J., Ouhon J.** and **Assoumou A. 2001.** Overall prevalence of orally transmitted parasitoses in children in Toumodi (Ivory Coast). Médecine d'Afrique Noire - 48 (10).
- **Agbere A. D.; Atakouma D. Y.; Balaka B.; Kessie K.; Kuakuvi N.; Gnamey D. K.; Assimadi J. K. 1995.** Digestive and urinary parasitism in children in a regional hospital in Togo: some epidemiological aspects. Médecine tropicale, vol. 55, n°1, pp. 65-67.

- **Agbere A. D.; Atakouma D. Y.; Balaka B.; Kessie K.; Kuakuvi N.; Gnamey D. K.; Assimadi J. K. 1995.** Digestive and urinary parasitism in children in a regional hospital in Togo: some epidemiological aspects. Médecine tropicale, vol. 55, n°1, pp. 65-67.

- **Ahn Y.H. and Choi H.C. 2004.** Municipal sludge management and disposal in South Korea: status and a new sustainable approach. Water Sci. Technol. 50 (9): 245-253.
- **Ait Melloul, Amahamid O., Hassani L. and Bouhoum K. 2002.** Health effect of human wastes use in agriculture in El Azouzia (the wastewater spreading area of Marrakech city, Morocco), International Journal of Environnemental Health Research 12, 17-23.
- **Alaoui AS., Khaloufi A., Boukachabine K., Agoumi A., Zouhdi M., Bourjouane M. and Alaoui MA. 1997.** Place of microbial and parasitic agents in infantile diarrhoea. Revue Marocaine de Biologie- Infesctiologie, 1997 III (3), 53-57.
- **Amahmid O., Asmama S. and Bouhoum K. 2002.** Urnab wastewater treatment in stabilization ponds: occurrence and removal of pathogens. Urban Water; 4: 255-262.
- **Aplogan A., Schneider D., Dyck J.L. and Berger J. 1990.** Digestive parasitoses in young children outside hospital. Ann. Pédiatr (Paris), 1990, 37, n° 10, 677-681.
- **Archambault A. 1997.** Fundamentals of gastroenterology: pathological

conditions and therapeutic approaches. Archambault A, Beaudry R, Paré P and Thompson ABR. Canadian Association of Gastroenterology, p. 747.

▪ **Ayadi A. Mahfoudh A. and Mahjoub F. 1991.** Intestinal parasitosis in children: Bilan de 2 ans dans le Centre Hospitalo-Universitaire de Sfax. Médecine d'Afrique Noire. 1991. 38 (8/9).

▪ **Bachta E., Zenaidi N., Belkaid M., Tabet-Derraz 0. 1990.** Assessment of intestinal parasitoses encountered in Algiers (1984-1988). Bull. Soc. Path. Ex. 1990; 83: 510-516.

▪ **Belghyti D., El Kharim K., Bachikh J. and Gabrion C. 1994.** Parasitological characterisation of waste water from Lake Fouarat (Kénitra - Morocco) and evaluation of the epidemiological level in a rural population in contact with this water. Proceedings of the Fourth International Conference of French-speaking Limnologists. Marrakech, 25 - 28 April 1994. Tome II.

▪ **Ben Said M., Messidi-Triki S., Afraoui N., Ben Rachid M.S. 1981.** Les parasitoses digestives de l'enfant. Revue Tunisienne de Pédiatrie. December 1981, n°4, PP 248-253.

▪ **Bigard MA 2001.** Oesophagus, stomach, duodenum, intestine, liver, biliary tract, pancreas, miscellaneous. Glaxo Wellcome Proceedings. Marly-le-Roi. p. 54.

▪ **Bigard MA 2001.** Oesophagus, stomach, duodenum, intestine, liver, biliary tract, pancreas, miscellaneous. Glaxo Wellcome Proceedings. Marly-le-Roi. p. 54.

▪ **Bouhoum K. 1996.** Etude épidémiologique des helminthiases intestinales chez les enfants de la zone d'épandage des eaux usées de Marrakech / Devenir des kystes de protozoaires et des œufs d'helminthes dans les différents systèmes extensifs de traitement des eaux usées. Doct. thesis, Univ. Marrakech. 227p.

▪ **Bouhoum K. and Schwartzbrod J. 1998.** Epidemiological study of intestinal helminthiasis in a Marrakech raw sewage spreading zgne Zeintralblatt Hygiene Umweltmed, 1998, 200, 553-561.

▪ **Bourré P. 1983.** Aide-Mémoire de Parasitologie et de Pathologie Tropicale, 3ème edition, Médecine-sciences, Flammarion, 289p.

▪ **Bretagne S., Jacovella J., Breuil J., Guillot F., Liance M. and Houin R. 1990.** Cyptosoridiosis in children: Epidemics and sporadic cases. Ann. Pediatr, 1990, 37, 381-386.

▪ **Buchy P. 2003.** Digestive parasitoses in the Mahajanga region, west coast of Madagascar. Bull. Sco. Path. Ex. 2003, 96, 1, 41-45.

▪ **Cevenini R., Varoli O., Rumpianesi F. and Mazzaracchio R- A. 1985.** two-year longitudinal study on the etiology of acute diarrhea in young children in Northeren italy. Microbiologica, 1985, 8, 51-58.

■ **Choukri M., Aoufl S., Douabi I., Agoumi A. 2007.** Intestinal parasitosis at Rabat University Hospital. Espérance médicale, 2007, vol. 14, no139, pp. 287-290.

■ **Cifuentes E., Bmlumenthal U., Ruiz-Palacios G., Bennet S. and Quigley M. 2000.** Health risk in agricultural villages practicing wastewater irrigation in central Mexico: perspectives for protection. 2000. Schrifter. Vet. Wasser Boden Lufthyg. 105: 249-256.

■ **Cisse AO, Ouattara A, Thellier M, Accocebery I, Biligui S, Minta D, Doumbo O, Desportes-Livage I, Thera MA, Danis M, and Datry A. 2002.** Evaluation of an immunofluorescent antibody test using monoclonal antibodies directed against Enterocytozoon bieneusi and Encephalitozoon intestinalis for diagnosis of intestinal microsporidiosis in Bamako (Mali). J Clin Microbiol 2002; 40 : 1715 -1718.

■ **Courouble G., Dufillot D., Sans A., Malpote E., Berchel C. and Nicolas M. 2000.** Survey of acute infantile gastroenteritis at the Pointe-à-Pitre/Abymes University Hospital, Guadaloupe. November 1997 to March 1998. Bull. Soc. Pathol. Excot. 2000, 93, 1, 58-61.

■ **Debonne JM and Rey P. 1998**. Diagnostic approach to acute abdominal pain. In: Gastro Entérologie 1: Démarches diagnostiques, Explorations - Nosographie 1. Debonne JM and Bernard JP. Doin éditeurs, p.15-35.

■ **Develoux M., Mouchet F. and Labo R. 1986.** Digestive parasitoses of schoolchildren in the Republic of Niger. Bull. Soc. Path. Ex. 79, 1986, 571-575.

■ **Dianou D., Poda J.N., Savadogo L.G., Sorgho H., Wango S.P. and Sondo B. 2004.** Intestinal parasitosis in the Sourou hydroagricultural complex in Burkina faso. 2004. VertigO - La revue en sciences de l'environnement, Vol 5, No 2, September 2004. 1-8.

■ **Diouf S., Diallo A., Camara B., Diagne I., Signate H. SY, Sarr M., Fall M. 2000.** Intestinal parasitosis of children in rural Senegal (Khombole). Médecine d'Afrique Noire: 2000, 47 (5).

■ **EHOLIE S. P., M. COULIBALY, O. KRA, D. ADOU BRYN , Y. Z. YAPI,**

E. AOUSSI, E. BISSAGNENE, A. KADIO. 1999. Diarrhoea associated with intestinal parasitosis in man (cote d'ivoire): epidemiological, clinical, aetiological and therapeutic aspects. Médecine d'Afrique Noire: 1999, 46 (1).

■ **El Kettani S. and Azzouzi E.-M. 2006.** Prevalence of helminths in a rural population using wastewater for agricultural purposes in Settat (Morocco). Cahiers Santé vol. 16, No. 4, October-November-December 2006.

■ **El Kettani S., Azzouzi E.-M. and Maata A. 2006.** Prevalence of Giardia intestinalis in a rural population using wastewater for agricultural purposes in

Settat, Morocco. 2006. Medicine and infectious diseases 36 (2006) 322-328.
▪ **Fraser G.C. and Cooke K.R. 1991.** Endemic Giardiasis and municipal water supply. Public health briefs, 1991. 81, 760-762.
▪ **Gbadoe A. D. and Koffi K.S. 2005.** Intestinal parasitosis in children under five years of age in Togo and IMCI recommendations. Archives de pédiatrie 12 (2005) 1400-1406.
▪ **Gendrel D. 1990.** Parasitic diarrhoea in children. Ann Pediatr 1990; 37: 590-593.
▪ **Gendrel D. 2003.** Parasitic diarrhoea in eutrophic and malnourished children. 2003. Med Trop 2003; 63: 442-448.
▪ **Gentilini M. and Duflo B. 1986.** The relationship between natural environments and communicable diseases. Relations entre les milieux humains et l'état sanitaire. Méd Trop, 1986, 5: 9-15.
▪ **Gilles HM. 1995.** Soil-transmitted helminths. In " COOK GC, ZUMLA A - Manson's Tropical Diseases, 20th edition". SAUNDERS ed, London, 1995, vol 1, pp 1369-1412.
▪ **Golvan Y-J. 1990.** Elément de parasitologie médicale. Flammarion médecine et sciences. 5ème Edition. Paris.
▪ **Guerrant RL and Bobak DA. 1991.** Bacterial and protozoal gastroenteritis. N Engl J Med, 325: 327-40.
▪ **Habbari Kh. 1992.** Impact de l'utilisation des eaux usées sur l'épidémiologie des helminthiases et de la croissance chez l'enfant d'El Azouzia. Thèse de 3ème cycle. Fac. Sci. de Marrakech.
▪ **Habbari Kh. 1996.** Impact de l'utilisation des eaux usées brutes sur l'épidémiologie du parasitisme intestinal et de l'état nutritionnel dans la province de Beni-Mellal, Maroc. PhD thesis. Fac. Sci. de Marrakech. 164p.
▪ **Habbari Kh., Tifnouti A., Bitton G. and Mandil A. 2000.** Parasitoses intestinales et pollution de l'environnement : cas de 1343 enfants à Beni-Mellal (Maroc). La Tunisie Médicale, revue mensuelle, volume 78, N° 2, 109-114p.
▪ **Hajfani M. 1976.** Assessment of the activities of the parasitology laboratory of the Rabat University Hospital. Thesis in Medicine. Rabat. 86p.
▪ **Hamada A. et al. 2003.** Changes in the prevalence of intestinal parasites among Japanese expatriates in developing coutries. Kansenshogaku Zasshi, 77 (3): 138-45.
▪ **Haslay H. and Leclerc C. 1993.** Microbiologie des eaux d'alimentation. Edition Tec et Doc-Lavoisier, Paris.
▪ **HUAET J-P. 2004.** Health and the environment. Editorial, Report, 417-419
▪ **Jombo G. T. A., Egah D. Z. and Akosu J. T. 2007.** Intestinal Parasitism,

Potable Water Availability and Methods of Sewage Disposal in Three Communities in Benue State, Nigeria: A Survey. Annals of African Medicine Vol. 6, No .1; 2007: 17 - 21.
▪ **Kennou M.F., Othman O., Maalej S. 1979.** Les parasitoses intestinales de l'enfant. Results of a coprological survey carried out in the Gabès region. Archs. Inst. Pasteur Tunis, 1979, 56 (3), PP 215-223.
▪ **Khaloufi A., Alaoui A.S., Tligui H., Chabaa L., Boukachabine K. and Agoumi A. 2000.** Parasitic and mycological epidemiological study in a Rabat crèche. Maroc Médical, tome 22, n°2, June 2000.
▪ **Kightlinger LK, Seed JR & Kightlinger MB. 1995.** The epidemiology of Ascaris lumbricoides, Trichuris trichiura and hookworm in children in the Ranomafana rainforest, Madagascar. J Parasitol, 1995, 81, 159-169.
▪ **King M. 1990.** Health is a sustainable state. Lancet 336 (8716): 664-667. Sep. 15, 1990.
▪ **Laamrani El Idrissi A., Lyagoubi M., Barkia A., Ayoujil M., Mahjour J. 1999.** Prevalence of intestinal parasitosis in three Moroccan provinces. Eastern Mediterranean Health Journal, 5(1): 86-102.
▪ **Lahlou T. 1983.** Bilan parasitaire dans la province de Kénitra (Maroc) établi d'après la pratique d'un laboratoire privé. Doctoral thesis in Pharmacy. University of Bordeaux II. France. 101p.
▪ **Lamghari Moubarrad F.Z. 2005.** Parasitological characterisation of El Jadida wastewater, its impact on the coastline (water and sediments) and on the child population in the discharge zone. PhD thesis. Fac. Sci. El Jadida. 168p.
▪ **Lariviere M. 1987.** Parasitologie tropicale: les grandes endémies. Epidemiology - Prophylaxis (medical and social professions). Ed. Marketing, 1987, 108-11 Paris.
▪ **Melyani M. 1983.** Enquête sur le parasitisme intestinal de l'enfant scolarisé au cours préparatoire à Sidi Yahia du Gharb en 1983. Thèse Médecine. Université Med V. Rabat. 40p.
▪ **Menan E.I.H., Rouamba E., Ouhon J., Nebavi N.G.F., Adjetey T.A.K., Barro-Kiki P.C. M.K.n Penali K.L. and Kone M. 1997.** Helminthiases **Results of five years of parasitic coprology at the Cocody Institute (Abidjan-Côte d'Ivoire). Médecine d'Afrique Noire. 44 (7).**
▪ **Minta DK, Dembélé M, Dolo A, Sidibé AT, Diarra AS, Konaté A, Diarra M, Diakité A, Sidibé AF, Traoré AK, Maiga MY, Pichard E, Traoré HA, Doumbo O. 2007.** Digestive parasitoses in patients infected with HIV/AIDS in the internal medicine and infectious diseases departments at the Hôpital du Point "G" Bamako - Mali. 2007. Mali Médical 2007 T XXII N° 1. 33- 36.

▪ **Moe C. 2001**. Effects assessment: determinants of risk and outcome of waterborne infections. In : Microbial pathogens and disinfection by-products in drinking water - Health effects and management of risks. Craun G, Hauchman F and Robinson D. ILSI Press, p. 233-58.

▪ **Ndir I., Gaye A., Sy M., Gaye O. and Ndir O. 2002.** Prevalence of intestinal parasitosis at the Roi Baudouin health centre in Guediawaye (Senegal). Dakmèd 2002, 47, 2.

▪ **Nozais J.-P. 1998.** Parasitic diseases and faecal peril: helminth diseases. Manuscript N° PF 03. Journée en hommage au professeur A. Dodin, Paris.

▪ **Omar M.S., Mahfouz A.A.R. and Moneim M.A.** 1995. The relationship of water sources and other determinants to prevalence of intestinal protozoal infections in a rural community of Saudi Arabia. Journal of Community Health. 1995. 20, 433-439.

▪ **WHO 1982.** Basic techniques for the medical laboratory. Etienne Lévy. Lambert.

▪ **WHO 1991.** Guidelines for the safe use of wastewater and excreta in agriculture and aquaculture. WHO/UNEP. Geneva. 205p.

▪ **WHO 1996.** Water and Sanitation: Aide Memoire No. 112, Press Releases. May 1996.

▪ **WHO 1997.** Analysis of wastewater for recycling in agriculture. Manual of laboratory techniques in parasitology and bacteriology, Geneva (1997) 31p.

▪ **WHO 2001.** Standard operating procedures in medical biology. 1ère Edition. Bureau des laboratoires.

▪ **WHO 2003.** Guidelines for safe recreational water environments. Volume 1: coastal and fresh water. Water sanitation and health, WHO, Switzerland.

▪ **WHO 2003.** Guidelines for safe recreational water environments. Volume 1: coastal and fresh water. Water sanitation and health, WHO, Switzerland.

▪ **Ouattara M., Dieudonné Silué K., N'Guéssan AN., Yapi A, Barbara M., Raso G., Utzinger J., N'Goran E. 2008.** Prevalences and polyparasitism of intestinal protozoa and spatial distribution

Entamoeba histolytica/Entamoeba dispar and Giardia intestinalis

in rural pupils in the Man region of Côte d'Ivoire. Cahiers d'études et de recherches francophones / Santé. V18, Issue 4, 215-22.

▪ **Pitisuttithum P., Migasena S. and Juntra A. 1990.** Socio-economic in Thai. Adult residing in and around Bangkok metropolis. J.Med.Assc. Thai. 1990, 73 (9), 522-534.

▪ **Rambaud JC. 2000.** Gastroenterology. Paris: Flamarion Médecine-Sciences, p. 580.

▪ **Schubair ME et al. 2000.** Intestinal parasites in relation haemoglobin level and nutritional status of school children in Gaza. Journal Egypt Soc Parasitol, 30 (2): 365-75.

▪ **Tligui H. and Agoumi A. 2006.** Prevalence of intestinal parasitic carriage among schoolchildren in Tiflet (Morocco). Revue Francophone des Laboratoires. 2006. (386) 65-68.

▪ **Tligui H., Oudghiri M., Bahji M., and Agoumi A. 2002.** Epidemiological and pathological data on Blastocystis hominis. Revue Marocaine de Biologie-Infestiologie. 2002. Tome X (1). 13-18.

▪ **Ulukanligil M., Seyrek A., Aslan G., Ozbilige H., Atay S. 2001.** Environmental pollution with soil transmitted helminths in Stanliurfa. Turky. Mem. Inst. Oswaldo Cruz, 96 (7): 903-9.

▪ **Vermeulen MS, Steyn NP, Nel J et al. 1997.** Socio-demographic, sanitation and hygiene factors related to nematode infection of a rural population in the Northern Province. South Afr J Epidemiol Inf, 1997, 12, 85-90.

▪ **Watkins WC, Cruz JR & Pollitt E. 1996.** The effects of deworming on indicators of school performance in Guatemala. Trans R Soc Trop Med Hyg, 1996, 90, 156-161.

▪ **Watkins WE & Pollitt E. 1996.** Effect of removing Ascaris on the growth of Guatemalan schoolchildren. Pediatrics, 1996, 97, 871-876.

VII. APPENDIX

Annex I: Questionnaire "epidemiological survey of intestinal parasitism in children hospitalised at the Kenitra Regional Hospital".

IDENTIFICATION

Card no

Date

Last name:

First name :

Province:

City:

Environment :

U R

Sex: M F

cm

Age :

Weight :

KgHeight :

School level:

Father's profession:

Mother's profession :

Number of people in the household:

HYGIENE

Habitat :

Hard

Slum

Other

Nbr. of pieces

Toilets : Yes No

AEP : Drinking water network

Well

Public fountain

Other : Spring Wadi

Seguia

Canal

Lake

Liquid sewage:

Sewage network

Septic tank

Open drain

Solid waste disposal:

Disposal of household waste

Public collection

Collective rubbish bin

Near the home

Hygiene personal : condition of hands and nails

Hand washing before meals:No / / **Yes (water) / / Yes (water + soap)** Yes (water)

Yes (water + soap)

CLINICAL INFORMATION

Digestive disorders:Yes //No/If yes, which ones:.

RESULTS

1. **Parasitological examination of stools**

Stool appearance		
Solid - moulded / /	Presence of pus	/ /
Soft/ /	Presence of mucus	/ /
Semi-liquids/ /	Presence of superimposed blood	/ /
Liquids/ /	Presence of mixed blood	/ /

Testing for parasites

Negative

Positive

*By direct examination with physiological water

**By direct examination with Lugol's

*By examination direct after concentration Method (s) of

Parasites found in stools

Haematology examination (see attached results sheet)

OTHER COMMENTS

SUMMARY

MIX
Papier aus verantwortungsvollen Quellen
Paper from responsible sources
FSC® C105338

Printed by Books on Demand GmbH, Norderstedt / Germany